I am Termite

Mike Pearce

DEDICATION

This book is dedicated to all those who worked with
me on termites in this country and overseas who
looked into the secret, wonderful, social life of
termites and the problems that they could cause.

.

CONTENTS

Quote

The Death of Solomon

King Solomon died silently holding his staff which supported him. The Jinns believed that they knew everything but they did not know Solomon was dead, but God did. After some days, God ordered a little worm of the earth (a termite) to eat away at his staff so that Solomon fell to the ground. This showed that the Jinns (along with humans) did not know everything but God did.

Based on the Quran Chapter 34 Verse 14

1 ONE OF THE ONE BORN EVERY SECOND

Imagine you are in a dark damp room. Right In the centre of a pitch black, cavern like clay cell, rather like a clay roasting pot lies my mother the queen. She had flown from her home nest with her partner many years ago, lost her wings and burrowed with her partner quickly into the rain sodden earth. Many hundreds of her fellow fliers had failed to find a mate and had either been eaten by hungry birds and lizards or attracted to street lights and lit windows where they floundered. She was now a princess, destined to remain under the earth, eventually many metres down and become queen; living in the dark and trapped forever. To survive the royal pair would need to feed on their own fat stores and protein which they would get from the breakdown of their wing muscles never to be used again. To help further they would have to eat their appendages linked to flight and wing pads as they were no longer needed. They mated after a few weeks and to continue her survival my mother would even have to eat some of her first eggs, thankfully not me. Her first born workers would then start to excavate the royal chamber and burrow to the outside to collect food, the start of our colony.

Both my mother and father had carried with them in their gut spores of a fungi without which the colony could not survive. We termites rely for food on the breakdown products of plant and wood material by fungi from food collected. The fungus would invade the first food brought down initially by my brothers and sisters so that they all could survive and feed future offspring. Eventually they would excavate special chambers for these combs and develop a royal chamber deep in the ground for my parents to live in.

As the colony numbers increased my mother's body would begin to swell eventually to become a transparent fat filled sausage sized sac with her little insect darkened cuticle segments spaced out along her length. Her little head and fore body which had not expanded would be left sticking out at the front. Her minute legs would no longer be able to move her along. She now would move just through contractions of her body. These body movements would help her breathe, sucking in air and expelling it through the air tubes leading into her body. Movement would also help her circulation. Her partner my father the king would still remain at her side tiny in comparison (one fifth the length) and would take a lot longer to walk around his huge egg laying machine wife than before. My mother's small hairs all over her body, once helping with her sensory perception, would now become now much reduced. Also both antennae would be partially chewed off as they will never need these totally again. They now remained in pitch blackness; their beautiful domed compound eyes which saw the world for a few hours also would now become redundant. Their freedom and exposure to light was no more.

Initially my mother would be attended to by her offspring. Both parents would be fed and cleaned by the youngest ones. My mother the queen would eventually be in control of thousands of her offspring sending chemical messages in her secretions and producing volatile scents (pheromones) through the colony.

My mother as she developed would increase her laying capacity to produce eventually up to one egg every second. She may in all produce 30-40,000 eggs per day. A super egg laying machine reaching up to 10 million per year. The importance of feeding the queen by her worker children therefore cannot be over emphasised. They would also have to constantly remove these eggs to other moist chambers to avoid massive overcrowding. Unlike us my mother can live for 15-20 years and in that time she will create a huge metropolis.

Her wet shiny, soft, slightly curved jelly bean shaped, clear bluish, hexagonally marked tinged eggs when produced are covered in an attractive liquid which our workers love to drink. Each egg is stuck to its neighbour by salivary secretions and placed close to fungus combs to keep moist.

 I was still in my egg shell. Workers seemed to know I was about to hatch and they had to tunnel through the egg pile to remove me. I struggled, kicking my legs, and freed myself by eating through the outer casing, helped by a worker who with others started to consume my egg shell. They then groomed me and cleaned me all over enjoying eating what was left of the egg contents. I had made it, 'I am now TERMITE'. Luckily I was not are eaten as some eggs are by workers while cleaning the queen. I now am sitting on top of a pile of eggs, like a small glass statue on a raised plinth in Trafalgar Square. I sat motionless. My sensory hairs were large compared to my size. A bit of nuisance but these were all I would need to help me in the darkness which was now my world. Eventually I was lifted up and taken out of the chamber by a young worker through one of the series of exit holes circling the royal cell to a large damp chamber piled high with eggs and next to something spectacular and hard to describe a fungus comb.

Many of our ancestors (lower termites) like some wood boring beetles picked up a range of small single cell organisms called protozoa. These live and thrive in their guts and produce enzymes which break down eaten wood to the sugars and chemicals needed for them to survive. Unfortunately my family does not depend greatly on these microorganisms but we eat wood and plant material which we chew and digest. We excrete this as small, soft, ball- like pellets. We deposit them while doing a sort of dance similar to bees, turning round several times to check where to place these pellets and checking what is produced. We then deposit the whole pellet by raising our middles and lowering our hind end and using contractions. Then with our mouthparts and using saliva we mould these pellets of semi digested food together to start to build a fungus comb. First we build a base then a vertical pillar which we join to another pillar to form an arch. By connecting these pillars and arches a wonderful labyrinth of tunnels and caverns are produced. These form the comb like structure for the fungus to live on. Both fungi and comb act as a food supply for the many thousands of us. We have often been referred to as gardeners as our workers have used spores of fungi originally carried by our king and queen in their guts to inoculate new areas of comb. These specific fungi grow deep into the cavernous comb and digest the wood and plant material in the pellets deposited there. The fungus then produces white threads called mycelia over and into the comb as well as producing small fungal balls attached to these. Without this digesting fungus on the comb we would all die and our workers have to continually add new material to the comb to continue this process.

My brothers and sisters can develop into what are called Major (large) workers or Minor workers (smaller). They vary in size and can have different roles within the colony. Both major and minor workers feed on comb nodules (the white fluffy fungal masses) where they can get the enzyme cellulase which breaks down cellulose in plants. Older workers eat specific areas of comb which are more attractive to them than others. On this comb next to me are scattered youngsters. I am now one of them. This close proximity to the food source means that I can be fed by other workers until ready to feed myself on the fungal comb. Some fungus combs are placed in many small chambers under the soil or may just be as a large comb centrally placed in the nest. I've found I can attract nurse termites by oscillating too and fro, showing agitation. Using this method I can solicit food from my fellow workers and make myself a real nuisance.

Both male and female sexes in our colony work together and are not separate like bees. But as with all termites we are subject to a caste system which relates to our specific roles in the colony. Through this we work together to support the whole colony to achieve our success and survival. We use trophyllaxis which is food exchange by mouth of semi digested material in our crops or anus secretions. We groom each other to request these secretions especially if there is a shortage of food or lack of moisture, then transfer is important. Our different roles can depend on age as well as sex. The younger and adult workers will help feed our young and also feed the royal family who are trapped away from the fields of fungus combs.

In this dark environment, sensory and chemical communication is important. Sensilla vary in number over our bodies. Numbers increase at the end of our appendages for contact and detection.

Chemicals are produced by my mother the queen through secretions passed out at her hind end to help control the colony. These messages travel throughout the colony then back to the queen so she can assess what is happening.

My mother, through chemical concentrations or information passed by sensory interactions and pheromones, is able to adjust the balance of castes in the colony and their behaviour. Our whole colony works as one unit, each member responsible for their specific tasks and ready to sacrifice their lives for the sake of the colony.

2 MY ROLE IS DETERMINED

I have remained in my youth in the nest helping to groom and clean workers. I also had the privilege of cleaning the queen and soldiers' antennae and legs. In doing this I would always groom the head and antennae first. If I groomed the palps around their mouth then I found they would often give me food or other secretions by mouth. We often took part in multiple grooming which was great fun and stimulated our sense organs. I was told not to do too much grooming as if I damage the body of a worker and cause the release of body fluids this could lead to cannibalism. This also made me wary when I was groomed by others, so I moved away if one of my nest mates got too carried away while grooming me.

At around twenty days I went outside our nest, which was about a metre high now, with other workers to forage for plant material. I would process this by chewing and passing it through my gut and adding the result to help build up the fungus combs and help feed some to the young.

For some reason my coordinated movements now were beginning to slow down inside the nest. This had already happened three times before after which I had shed my old skin. It was another moult on its way. Many soldiers had been killed in an attack by ants before I was born I was told. I really began to feel most peculiar this time. I lost my old skin which was removed by the help of fellow workers who enjoyed eating it, which is one of the perks of being a termite.

Only a few soldiers remained in the colony and so the level of Juvenile hormone had decreased. If the levels in the surrounding air in the colony were higher this would have inhibited new soldier development. This very low level of Juvenile hormone meant that there was a need for more soldiers and I was to become one of them, in fact a major soldier. This also meant that at this level of hormone new winged forms would now be produced in the colony ready to fly out into the outside world.

I remained immobile after my moult completely white without pigmentation, just like a shiny unglazed porcelain figure. Even my jaws were white. My role now was no longer as a worker, building up the combs and feeding others and feeding on the comb. It was now to be only for protection of the colony and my survival would depend on being fed by others as my jaws were no longer small and designed for cutting. They were now huge and long ready to grab and dispose of enemies. However, I would still be able to get dissolved sugars and water by sucking on the comb itself. My head was hugely enlarged, destined to become a lovely chestnut brown. I was also told I was of male sex, as I had two protrusions called styles at the back of my abdomen between two other protrusions called cerci. My eyes were not much use compared to the winged forms, but I still could distinguish movement. I was ready to defend the colony from invaders at all costs and protect our workers during foraging for food and building and repair.

3 BUILDING BEYOND BELIEF

We have always created our own environment by building castle like mounds and tunnelling right down to the water table. Down here the soil is wet and great for building our impressive towers and cooling us down. Worker routine is an obsession rather like OCD. Each task is set in stages and in sequence. If one part of this routine is interrupted then the workers may have to start again from the beginning. Mechanoreceptors i.e. sense organs responsible for detecting movement and contact, respond to stress on our bodies. Our palps (feelers) around our mouths using these mechanoreceptors ensure that there is fine monitoring on things we carry. Delicate items such as eggs and larvae are lightly held by our workers whereas soil particles require a stronger hold. Not all of us termite species build mounds and many have exploited other ways which bring them closer to their food sources.

Some of our relatives who are wood borers use these sense organs to detect lateral stress which is especially useful when burrowing in wood. This ensures that there remains a fine outer layer of wood when they burrow and feed near the outside part of the wood.

Without access to water our building activities would be very restricted, if not at all. We therefore make good use of wet soil after rain and also at night when the air is cooler and less water is lost from our building materials. Without moist soil we also would not be able to seal off exit holes if our nest was broken and our enemies would move in. I have in my mouth, like our workers, a structure called a hypophaynx which is used to suck water out of wet soil and water found on the fungus combs. This water can contain minerals and nutrients to help our survival. Some of us termites also have salivary reservoirs in our abdomen to store water. These may make parts of our body look swollen and transparent like little water containers. We need water also to regulate our nest temperature and may, in some cases, even have to burrow down to as much as 10 metres in order to reach the water table under the soil.

As with constructing fungus combs there is method in our building activities and a building sequence is essential. Some parts of the nest need to be stronger and thicker than the rest. An example of this is in the chamber we build around our parental royal pair. This needs to be really hard and thick to prevent enemies entering and also to provide a constant environment. I remain on guard while a few workers get together to start building using moistened soil carried in their mouths. They mould these building blocks into shape using their mouth parts. At a certain stage of construction, when these small pillars of soil reach a certain height, there is a trigger for the production of arches which extend into tunnels. Our direction of building can be influenced by the concentration of pheromones. As our colony grows there is a need for more or larger chambers. Soil has to be taken to the outside to provide more room for fungus combs, the growing numbers of termites and the need for improved ventilation.

Our fungus combs break down organic matter and like a compost heap, during this process produce a lot of heat. Our outside environment in the day can also be hot. To contend with this we need to bring in cooler air from near the base and release this nearer the top so often we leave holes near the sides and base and have a central tall chimney at the top for release of the hot air. As these are potentially entry holes for invaders we, as soldiers, need to ensure they are guarded well.

For such a small creature, we are some of the world's greatest architects. Our soil mounds can reach 12 metres across and over 4.5 metres high. Some buildings constructed by humans have used similar architecture using similar ventilation systems to us. In some cases we can dig out large cavities and produce a mushroom like platform deep down inside the nest with downward facing vanes circling around the underside. This may also help with air circulation.

As well as building nests we can tunnel underground and emerge at the surface to collect food. If there is little vegetation around our nest and little shade, we need to build soil runways above ground to protect us from the heat of the day and from predators when we go out to forage for food. Humidity receptors are present on our antennae.

When building we are exposed to predators so, as a soldier, I have to accompany the workers building runways and passing through these runways. Larger (Major workers) are often used for gallery expansion and small workers for repair of these runways. Where a good source of food is found or when we travel up larger food sources such as trees we may build large canopy like nests which are sheets of soil to make maximum use of the resource. These soil shelters are more brittle and often need more repair. They provide a large area for predators to enter so they need soldier presence.

Our building and activity often has cycles throughout the year. In the rainy season our runways and nests can often be eroded and need repair. They can often become flooded in some regions or affected by irrigation of crops by farmers. With increased vegetation growth or new sources of food available our activities are increased.

Clay from deeper in the soil forms an important part in helping to increase the strength of our mounds and runways. Unfortunately, humans have found a use for this. Our mounds may be reduced, but not devastated, by the removal of clay. Clay is used for making bricks, and plastering floors and walls of mud huts and houses. This clay is also used for making ovens and pots.

4 OTHER TERMITE HOMES

There exists a huge range of soil nests. Some are mushroom shaped or domed others are flattened like grave stones above ground.

Some of our other relatives spend their lives in wood or exist as small colonies growing next to or inside wood or trees. They tunnel out the inside and fill the gaps or line their tunnels with their faeces. If the wood is also being used by humans then they become pest species. Often as we are eating keep a thin layer of wood on the outside and even sometimes build pillars from our faeces to support these. This means the only way humans can detect our presence is when it's too late as floors and beams collapse or trees start to die. Wooden structures such as telegraph poles and fences can be eaten away by us and collapse. As pests we can cost humans millions of pounds for wood replacements or control methods. Human past history is often eroded by us termites as we can destroy books, manuscripts and historic buildings or furniture.

Others can make nests high up in trees made out of paper like carton. Some of these live in shaded rainforest and do not have runways but can forage in the open moving in long columns like ants. Their eyesight can be better developed than those foraging in the dark.

5 FORAGING

The more primitive kinds of termites have small, single cell organisms such as protozoa and bacteria in their guts. The protozoa have enzymes that can break down chewed wood for useable food inside their guts. Here there is no need for fungus combs. We fungus growers also can have bacteria in our guts which help aid with digestion but do not have protozoa. The fungus combs help provide our colony with food when there are no crops around or wood is scarce.

Some plants we do not like to eat as they can contain chemicals. However if faced with one of these as the only food source, we can become accustomed to these chemicals. An example is the creosote bush. What is useful is that often these repellent chemicals are on the inside of some trees so that we can just feed on the outer bark.

Our taste receptors are on our antennae, mouth and the flap over our mouth the labrum, plus other parts of our body even on our feet. Once we come into contact with cellulose this attracts us and we begin to feed. They call the cellulose an arrestant.

Chemical messengers (pheromones) are triggers for how we behave. They are given off as a scent or in body fluids from glands. They affect our development and social behaviour and help us to work together and receive instructions. Our queen, our mother, sends us instructions using pheromones. Different concentrations can have different effects. We have pheromones released by our sternal gland under our bodies used for trails. This gland is on the anterior part of the fifth sternite on our hind body segments. With this pheromone we can recruit more individuals to a food source and also more of us soldiers when needed.

We can feed in groups or circles helped by an aggregation pheromone from our labial glands and can make tunnels into food in different directions. As our colony grows we have to do more foraging to meet everyone's needs and build up our fungus combs. Sometimes our fungus combs become redundant and we build new ones to replace them. The fungus comb breaks down our collected food and we get our nitrogen from the fungi on the comb and also when we eat animal dung. We can store uric acid in our bodies. If we are unhealthy you may see this as white patches inside our bodies. If desperate or through over grooming we damage a colleague, we may end up eating them. This is another source of nitrogen.

Apart from wood and plant materials, especially cultivated crops and grasses, we love dung, especially from those animals where undigested material passes through them. Humans also scatter a lot of rubbish especially paper items which we can use as food and in some countries mats and house screens and straw roofs woven out of plants are a real luxury. Books in libraries can be attacked. Railway sleepers and even power cables have been eaten. Some termites though are purely soil eaters getting their nutrients from any organic material present in the soil.

We always have some soldiers, such as myself, present when workers are foraging. This means workers have to make our soil tunnels larger so that our soldier heads can fit through. Companiform sensilla regulate the opening and size. If desperate and we need food fast we can forage in open air.

We build foraging tunnels out of soil. This requires a lot of moisture to stick them together so that we are glad when it rains or water is dropped onto the soil. Our tunnels are very fragile as they are often temporary structures and they often collapse or are damaged. They offer protection from enemies as well as providing a quick route to the food sources. They also keep us out of the sunlight which would cause us to dehydrate. If sand is incorporated into runways then they are more fragile, the more clay the better. We can cover plants with this soil sheeting or the sides of large trees where they provide a good food source. The stronger the pheromone trails laid down the wider the galleries that are built and the more workers recruited for repair.

Older trees tend to have cracks in them or stumps of broken branches where we can enter. Once a good food source is found, we can forage for over 50 metres but this depends on the amount of use of the galleries and on the soil type and obstacles in our path. Not finding food means we have to forage wider and more randomly until we find some. We may even have to tunnel deeper into the soil to search out roots etc. Some decaying roots are attractive to us as they give off chemicals.

If our major workers find food they will recruit more major workers and also minor ones to transfer food back to the nest. Our major workers are the best cutters of plant material. We soldiers are not able to cut anything at all except predators possibly. During the day in very hot sun, plants can become stressed due to water loss and wilt. This is a good time to attack them as any defensive chemicals are at a lower concentration. The plants will have recovered at night. Often farmers who only have poor soils, or where water is limited, have crops which are often more subject to our attack. Attack on crops may be patchy where different conditions exist within an area of land such as stoniness, compaction or waterlogging. Often a farmer may not know of our existence as we travel up the roots into the plant or feed on tubers or pods under the ground. Our presence can make a difference to yield, as well as to saleability based on the appearance of the crop, for example scarring or holes in root crops such as yams or sweet potato. Humans need to investigate the abundance and locality of termite nests before planting crops or building structures using wood. With buildings, when built, they need to look at water runoff which could attract us termites. When tunnelling in wood it is always easier for us to eat through with the grain of the wood. You can always tell if the wood has been attacked especially by our dry wood termite friends as we expel as faeces small seed like pellets (frass) to the outside. These are ridged and extremely dry as our bodies have sucked out all the moisture in them before they are released.

The degree of attack can be affected by the hardness of wood and the amount of lignin present. Hardwoods tend to be more resistant than soft woods. Sometimes some parts of a plant may be more resistant than other parts. Often we may form a ring of attack around a woody plant.

This can cause water loss and eventually death which makes them more susceptible to damage by us. Fast growing trees (especially spring growth) are better and easier for us to manage as they have larger cells, thin walls and fewer fibres so that we can bite through these more quickly. New sapwood also has more sugar or starch stored than older heart wood. Whole trees can be eaten out as well as their fruits This is especially important to humans if they are a cash crop such as dates which have a good supply of sugar. Sugar is a bonus for us, especially in crops such as sugar cane and maize, the content depending on the age of the plant. This can affect the time when most damage occurs. Damage is often greater at the ripening stage. Our very existence depends on the environment and the seasonal changes that can occur.

A rare form of food source is provided where humans put pieces of wood into the sides of our mound. These act as an oracle for decision making depending on whether the wood is eaten or not by us.

6 BRAVING THE ELEMENTS

Water is a life saver but can also be a death sentence. Where flooding or irrigation occurs we can be trapped inside the nests or drowned while foraging. We may be lucky enough to rest in an air pocket in a runway and survive. A lot also depends on drainage and elevation, direction of flow and the height of the water table. If our royal pair are drowned then that is the end of our colony. Where seasonal floods occur or constant irrigation, many of us termites have adapted to these and after the floods flourish in large numbers. The time of submergence is important for survival. We can survive for a short while submerged, especially if in an air pocket in the mound. Prolonged standing water may not allow us to return to our nests in order to get food so we can die. Sudden flash floods can be a problem, washing away our soil and debris. On the other hand, as a bonus, flooding may bring with it plant material and wood to an area where much of the food available had been consumed already. Flooding may also kill crops which can then be attacked. Floods also encourage more growth of vegetation. High rainfall can ruin our soil mounds, washing down our chimneys, and covering our ventilation holes with runoff soil. Rain can also destroy our soil runways inside or out from the nest so that we have to repair these in order to maintain the flow of air through our colonies.

Near the coast in hot areas our only good source of moisture may be the morning sea fogs that move in and deposit moisture onto the soil and plants. Plants can be dripping with moisture and our soil runways get really wet so can be entered by predators. But these are easy to repair. We like these mists as we can actually drink this surface water, build more runways and take water back to our nest.

Temperature can also be a problem especially as we have soft abdomens with a thin cover (cuticle). Extremes of temperature, especially for long periods without rain can cause soil to dry out. This can be a problem especially where mounds are small and heat up quickly. Fungus combs involved in organic matter breakdown like a compost heap can heat up air in chambers in the mound. Even in the shade these can get to 29-30 degrees centigrade. We termites have learnt to open and close exit holes in chimneys to regulate ventilation and temperature. Temperature also affects foraging especially in open areas and in very dry areas such as desert where high sand temperatures occur. We mainly confine outside activity to the evening and early morning or in places of shade when temperatures are lower. Very low temperatures at night also slow down our progress but are not as serious as the high temperatures.

Dust or sandstorms can sandblast our mounds as well as cover our nests. Fire is devastating for those termites living in wood or having nests high up in the trees. Our mounds are also hardened by fire and smoke can harm us.

We as termites, are important for helping to break down organic matter and help aerate the soil just as earthworms do in temperate regions. By removing and burning forest

(slash and burn) and destroying our nests these lands
eventually become non-productive over several years of
farming.

7 ALWAYS DEFENDING

We are termites, not beetles or cockroaches. We may have hard heads and thoraxes but the hind part of us is soft tissue easily punctured or grasped by something that would like to eat us.

Our workers have strong biting mandibles that can bite off wood fibres. They can bite predators but many other predators such as ants also have strong jaws. We can detect foreign invaders by smell or even sound. Workers can tap the wood or mound to warn others but soldiers are more effective at this.

Major and minor soldiers are designed to dispatch prey. There is a wide range of mandibles with different uses. Some soldiers have powerful, long, sharp teeth which chop the predators in half. Others have strong, thick mandibles for crushing or sharp for piercing. There are even mandibles which, when open, are fastened back and are released suddenly, so as to flick the predators away. Others may have serrated saw shaped mandibles. Having large mandibles, however, means they cannot easily hold small predators so workers have to help out. I only have strong mandibles with teeth for cutting.

Some other species of soldiers have openings on the top of their heads or have long extended noses where they fire out toxins and glues at the end from their frontal glands or from salivary reservoirs. Major soldiers can hold many hundreds of times more white sticky secretions (often containing oils and quinones) than minor soldiers, so are more effective.

In some termites, setae (small hairs) or bristles act as paint brushes which are used to daub intruders' mouthparts with these toxins and sticky materials. Some other termite workers method of defence is by bursting open the hind parts of the abdomen showering the invader with their gut contents. Once this is done they have sacrificed their lives for the colony.

We soldiers come out at cool times of the day or at night to sit on guard while workers repair or build their nests. You can see us sitting, our antennae moving tasting the air. Workers come out and feed us as we can be on guard for many hours.

If we are attacked us soldiers stand at the front line busy fighting an advancing enemy, while the workers can seal off the entrances behind with wet soil, so as to form an impenetrable wall. It is therefore vitally important to have a good number of soldiers present for the defence of the colony. We sacrifice ourselves to the oncoming enemy as there is no turning back especially if our entrance hole has been sealed up. Major workers can also put up a great fight and can be more aggressive than us soldiers, their jaws being strong and with side teeth which can cut into predators. In wood dwelling termites some soldiers have plug like heads flattened in the front which just fits inside the tunnels forming a barrier to oncoming invaders. Our queen chamber has regular holes across the middle of its hard clay walls. This means each hole can be defended rather like slit windows in a castle, the archers now being the soldiers.

Below the knee in us termites is something called a subgenual organ. This has hairs (cilia) sensitive to vibration. These will signal the arrival of a predator so that termites move back into the centre of the nest and soldiers move out towards the outside. We can also release alarm pheromone which will trigger   tapping on the mound structure by workers or soldiers. For termites living inside wood this is even more effective as wood is a good transmitter of sound.

As we live in darkness we need to be able to recognise each other in our nests and tunnels and distinguish from intruders whether it be termites from another colony or other insects such as ants. Our pores on our outer cuticle secrete hydrocarbons in their waxes. These help us identify our nest mates. If an odour is picked up by our antennae that does not match then attack procedure and alarm is put into place. To help prevent attack we may remove soil from another part of the nest or runways. Minor workers also will begin spontaneous emergency construction to block off tunnels while we soldiers stand on guard.

A raid by ants can cause high mortality, but with a high rate of eggs produced by our mother queen this may not pose a serious loss except in young colonies. Predators like ants together with dead termites can be sealed off or buried with our same building soil. Nothing is wasted in termite colonies. If one of us bleeds and haemolymph release occurs from this injury, this may cause cannibalism.

Several beetles and fly larvae have become adapted to look similar to us termites. They also can produce attractive secretions. These intruders can be a problem as they may eat us or feed on our stored food. Assassin bugs are known to catch us for food.

Many animals can rest, make nests or take shelter in our nest. These often pick off some of our workers which venture outside the nest to forage or repair. Lizards and birds are always on the outlook for stray termites, especially our winged ones (alates) which form a great feast for many species.

On the other hand some animals such as anteaters use their curved strong claws to burrow deep into our nests to find us. They lick us up with their long sticky tongues. Aardvarks also use large claws and long tongues and have even the capacity to detect on which side of the nest we termites are most active, by detecting temperature difference.

Monkeys such as chimpanzees, to our horror, can push grass straws down our tunnels. This initiates attack by our workers and my soldier friends which bite the straw. The straw is withdrawn with us clinging to this. They then pass the straw through their lips to dislodge us into their mouths.
In some areas our mounds act as rubbing posts for large animals. Animals such as elephants can use our mounds to massage themselves, dislodging large amounts of soil. If the mound becomes a daily scratching post this habit can be a problem.

Our sausage sized mother, the queen, is considered a real delicacy to some humans and is even thought to be an aphrodisiac by some groups. Removal of the queen involves hard work. Humans have to dig deep down under the soil to get at the centre of the mound to find the royal chamber. We can in some cases if there is time, open up the royal chamber and smuggle her out. More often or not though she is captured, leaving our poor father the king queenless, lost and neglected because of his small size. This usually is the end of the colony, the rest of us termites dying off or being carried away by predators. On rare occasions we may survive this especially if our nest contains future queens (the alates) or contained another queen as sometimes happens.

We have many parasites and predators all around us. Our fungus combs are antibacterial as are the secretions from our wax glands on our bodies which protect us from pathogens such as ectoparasitic fungi. Some fungi may attach to the outside of us and grow a long stalk out of our bodies releasing spores into the air. Mites can creep onto and attach to our bodies especially when our colony is not that healthy. If pathogens or parasites attack us as we are a social community these could spread rapidly throughout our nest. We can be infected by small worms (nematodes), which if eaten can develop inside us and multiply. If we are infected or die we can be buried in soil and in some cases we are removed from the nest and dumped in a termite cemetery outside the nest. This is important as decaying termites can spread unwanted fungi and bacteria to our colony. In dry wood termites a sick or infected termite is either coated with hard secretions like a mummy or else sealed off in a gallery at both ends by this sealant so that

27

none of the colony can come into contact with it and spread the infection.

8 PRODUCING FUTURE KINGS AND QUEENS

The anal fluid of our queen had maintained a level in juvenile hormone that prevented new kings and queens (alates) from being produced. But now the level of hormone had fallen and a conversion to winged forms in our colony was in progress. They remain in their white, unglazed, porcelain like appearance until they become tanned to a chestnut brown and their wing cases are removed and eaten. The new generation of alates, before they leave, have to get fungus spores into their gut and this fungus is specific to our colony. They have developed large eyes and if they are the ones that fly in the dark then they tend to have larger eyes. They also, like other alates, have a pair of light detecting ocelli (eye spots like spiders) above their eyes. Sticky pads are also found on the bottom of their feet allowing them to climb up slippery surfaces.

Primitive termites are lucky in that if the queen or king dies one can get new queens and kings developing from workers. This is useful as a colony in a piece of wood may be divided if a piece of wood is sawn or broken into pieces. The piece or pieces of wood without a royal pair can develop their own and survive to produce young. A kind of budding off of a colony.

Our future kings and queens (alates) congregate, tightly packed in their pre-flight chambers, well away from the royal pair. Falling rain drops can be the trigger for flight and they leave from slits or holes in wood or flight tunnels in the mound. Soldiers are on guard while workers open up holes and herd out these winged insects for their quick escape. Often there are more males than females flying and you can have flights over several months. Our reproductives don't fly far as their wings which are similar sizes are not developed for this.

Not all prospective kings and queens make it to start a colony. Often fifty percent are lost at the time of flight. Correct timing is essential for all colonies so that termites can meet others from different colonies of the same species. Hundreds leave the nest, often after a few days of rain. They can though fly in the rain. Rain lowers the temperature.

At the ground the royal pairs can burrow quickly down under the surface. The release of large numbers ensures that predators become engorged i.e. full up so do not chase after all of us. Alates are photopositive, and are especially attracted to sunlight or at night to the full moon. Unfortunately any artificial bright light is also attractive. Many fly towards street lamps only to fall to find the ground is concrete. They have no soil to dig down into. To lose their wings they twist their abdomen or rub against something. If that doesn't work they are sometimes chewed off. Female alates have a tergal gland for pheromone to attract males. Once landed the female's sternal glands lays down scent for the male to follow in tandem and not get lost. Often one can see a female being chased by numerous males. The flight exit hole in the mound will be quickly sealed after their release by workers as predators may be abundant and be able to access the nest through these holes. It is important for all the winged forms to all leave. If they do not leave for some reason or all do not leave or have a defect then they can be attacked by our workers who will eat them for their fat and protein.

Many other insects lurk around lights. Examples include velvet mites, dragonflies, birds, toads, lizards, ants, mantis and spiders. Even bats hang around to feed on us.

Some of our alates may fly towards car lights only to be squashed under traffic wheels on the roads. The lights of houses also are very attractive and sometimes house porches are covered in hundreds of wings. If humans leave the windows or doors open they can find the room full of our royal pairs flying round the lights and racing around on the floor. If these alates fly in the day they are in even more danger as they provide an important food source for birds who catch them flying in the sky or when they fall to the ground. Often birds time their production of young with the season and month our alate flights.

Humans too are even more powerful predators. The winged form of us termites can be half protein, half fat. Our prospective kings and queens can be eaten raw, fried roasted in oil, or fried and ground to a powder which is added to food as a paste. They also can provide food for animals such as chickens or fish. Upturned pots or traps made out of sticks and leaves or animal skins or plastic sheeting can be placed over flight exit holes so flying termites can be collected from these. Water can be added to soil to simulate rain, or mounds can be beaten like drums to attract alates out. Torches are used at night to attract them to a particular spot. Villagers may own our mounds and these provide a reliable food source and clay source handed down to generations of families.

9 THEY TRIED TO CONTROL US

Eating crops or wood makes us a pest, something that must be killed or controlled. Numerous different ways have been tried but we have been around a long time and our existence is indicative of our success.

Chemical methods have been tried, whether instant effect or delayed. They put it in the soil with plants or on baits which we eat. They have even tried fungicides on baits to kill our fungi which we depend on or used chemicals to kill the wood digesting organisms in our guts. They spray wood, fumigate us under gas filled tents, and can use dusts to destroy our waterproofing waxes on our bodies. We know if our fellow nest members are being killed or affected and will act appropriately.

We have learnt to stay away or seal off the infected or poisoned individuals. Methods of biological control have included viruses, nematodes and fungi to infect us, but they have not been successful. Because of this it's a hard choice to isolate and reject infected individuals, but we are a social unit and any threat has to be dealt with straight away.

Repellents can be effective. Initially they tried extracts from leaves such as those from neem trees and even used dead animals buried or salt in the soil. Changes in design in buildings have been used to lower the humidity in order to make for example loft spaces less attractive. Also raising wooden structures up away from the soil surface and barriers (chemical or physical) placed under houses and around houses are used. Every day people will brush down our runways to prevent us entering buildings but we look for other pathways. Removing waste, plants or other materials is another method used so as not to attract us. However placing attractive food, such as piles of wood, rubbish or non-commercial crops a short distance away from crops or buildings is believed to divert our attention, and provide a centre for our control. Often little do they realise that more food helps us build up our colonies so we become more of a threat.

With the threat from dispersal of hundreds if not thousands of our winged forms and consequent new colony formation, the battle against us termites will always remain. Humans may always have to share the harvest with us termites a yearly variable cycle that extends back thousands of years.

10 STORED FOR ETERNITY

Workers only live a few months. Fortunately there is a high production of eggs to replace them. Some humans (especially termitologists) are collectors of termites. Even Charles Darwin collected them. As with many specimens they are kept locked away in the museums around the world such as the Natural History Museum in London. You will not find me pinned to a piece of cork like butterflies and beetles as my abdomen would just shrivel up and not good to look at. I am there together with some minor soldiers, major and minor workers and alates from my colony kept in small tubes of 80 percent alcohol with a label. They did not manage to get our queen but they do have other large sausage shaped queens which are stuffed into cylindrical tubes in the collection. Still today my colony continues to live out in Africa building up huge numbers and huge mounds. Around us there are now are many other colonies established after years of flights of alates from our mound, all brothers and sisters of this social network. But as for me I am pickled, destined to remain in alcohol for the rest of my days and remain eternal as long as the museum exists to be admired by many.

I am and will remain TERMITE.

ABOUT THE AUTHOR

Dr Mike Pearce is a scientist interested in animal behavior. He worked on biology and control of termites for over 21 years both in this country and overseas.

He has also worked as a lecturer in Health and Social Care at a college in Canterbury, Kent.

His aim in this book is to introduce people to the world of termites showing how they live and what can threaten their very existence.